GEESE FARMING FOR

BEGINNER

A   Guide To Raising, Breeding,   Caring For
Pochard - Tips On Housing, Feeding, Health
Management, And Profitability

Holden bodhi

Contents

DISCLAIMER

The information provided in this book, is intended for educational and informational purposes only. The content is based on research, personal experiences, and general knowledge about farming. It is not intended to substitute professional advice or expert consultation. Readers are encouraged to seek professional guidance when implementing any practices or techniques discussed in this book.

The author and publisher make no representations or warranties of any kind regarding the accuracy, applicability, or completeness of the contents of this book. Any reliance you place on such information is strictly at your own risk. The author and publisher shall not be held liable for any damages, losses, or injuries resulting from the use of the information provided.

Additionally, the author does not endorse, recommend, or affiliate with any individual, product, service, website, organization, or brand mentioned or referenced in this book. Any such references are solely for informational purposes, and no warranty or guarantee is implied. The inclusion of these references does not imply any endorsement or partnership by the author.

By reading this book, you acknowledge and accept that the author and publisher are not responsible for any consequences arising from your use of the information provided

CHAPTER ONE

An Overview of Farming Geese

The technique of breeding geese for meat, eggs, feathers, and even as guard animals or pest controllers is known as "geese farming." Geese are resilient birds that do well in a variety of environments and farming practices. Geese provide a unique mix of goods, including premium meat, nutrient-rich eggs, and down feathers valued for their insulating qualities, in contrast to chickens, who are more often farmed for meat and eggs.

Geese are often divided into many breeds, each distinguished by distinctive characteristics. Some are prized for their ability to lay eggs or as ornaments, while others are better suited for producing meat. The Embden, Toulouse, Chinese, and Pilgrim geese are popular breeds for farming, and each has unique traits related to production, disposition, and pace of development.

Goose production extends beyond agriculture; they are involved in environmental management. Because they are grazers by nature, they may help suppress weeds and increase soil fertility by eating grass and other plants. This means that raising geese is a more ecologically beneficial way to manage land since it uses fewer pesticides and promotes a more sustainable agricultural method.

Advantages Of Having Geese

Beginner farmers find raising geese to be an enticing option due to its many advantages:

1. Minimal Upkeep

When it comes to birds of prey, geese need less care than other species. They need less regular veterinarian care since they are resilient and resistant to many common poultry illnesses. When given the right conditions, geese may survive on a natural diet of grains and grasses, which reduces the need for feed. In addition, they are a sensible option for farmers in a variety of regions due to their robust immune systems and ability to withstand severe weather.

2. Adaptable Items

Geese offers a large selection of goods that are suitable for personal use or for monetization. Their flesh is tasty, lean, and prized in many cultures for its delicacy.

Despite being less frequent than chicken eggs, goose eggs are prized by bakers and chefs for their unique recipes because they are bigger, more nutrient-dense, and have a higher quality.

Furthermore, goose feathers—particularly down feathers—are used to produce high-end outdoor clothing and bedding, which generates extra income.

3. Land management and pest control

Geese make great foragers and natural grazers. They are the perfect way to manage weeds in agricultural settings since they eat a lot of grass and other vegetation. They may enhance the general health of the soil and lessen the demand for herbicides by grazing. Geese are often used as live lawnmowers in orchards, vineyards, and even organic farms. This allows them to manage the development of undesired vegetation without endangering crops. Their importance as integral parts of sustainable agricultural operations is further enhanced by their capacity to eradicate weeds from the land and enrich it with their excrement.

4. Protective Species

The protective character of geese is another advantage of having them. Geese are renowned for their territorial nature and acute danger awareness. When they smell danger, they let out a loud honk to warn farmers about potential intruders or predators. Because of this, they make good guard animals that can ward off small predators like foxes and raccoons and even act as a warning system for potential dangers.

5. Sustainability of the Environment

The raising of geese is a sustainable practice. By feeding on plants, geese aid in the natural management of land by reducing soil erosion and fostering biodiversity. Furthermore, compared to bigger animals, geese emit fewer greenhouse emissions, and their dung is an excellent source of organic fertilizer that improves soil quality. Consequently, regenerative agriculture systems that seek to improve and repair ecosystems may include goose farming.

Important Things To Know For Novices

There are a few crucial things to think about before beginning a geese farm in order to guarantee a successful and satisfying venture:

1. Space Needs

Since geese are an active bird that likes to graze and forage, they need enough room to live well. Their health and well-being depend on having access to a large, open space that includes grass or pasture. Since geese are excellent swimmers, having access to water, such as a pond or stream, may also be beneficial to them. Beginners should make sure to provide enough room on their farm for both grazing and nesting.

2. Appropriate Housing

Even while geese are resilient and can withstand a variety of weather conditions, they still need protection to keep out predators, strong winds, and freezing temperatures.

Enough shelter may be obtained with a basic, well-ventilated shed or barn, particularly in the winter. There should be enough space in the shelter for the geese to walk about comfortably and deposit their eggs. Safeguarding the shelter against foxes, weasels, and big birds of prey is also crucial.

3. Nutrition and Feeding

Since their primary food is grass, geese may be raised cheaply in regions with an abundance of pasture. However, during the winter months when grass may be sparse, their nutrition should be supplemented with grains, vegetables, and specially-made poultry feed. For their health, they also need access to clean, fresh water, especially if they don't have access to a natural water source.

4. Breeding and the Production of Eggs

It's crucial for anybody looking to grow their flock of geese to comprehend egg production and breeding. Geese are migratory breeders that produce eggs in the spring; they usually pair for life. Beginners should set up a proper habitat for egg laying and hatching and get acquainted with the unique breeding behaviors of their selected breed. Nest boxes that are suitable and a peaceful, uninhabited place can help geese successfully lay eggs and raise their young.

5. Prevention of Diseases

Even though geese are hardy birds, disease outbreaks may still be avoided by maintaining proper cleanliness and biosecurity measures. Keeping an eye on their health, cleaning their shelter on a regular basis, and giving them clean water may all help to keep the flock disease-free. To stop the transmission of disease, fresh birds should also be housed separately before being added to an established flock.

Natural land management, low-maintenance care, adaptable goods, and environmental sustainability are just a few advantages of raising geese. To guarantee a successful endeavor, newcomers should be aware of the needs for space, housing, feeding, breeding, and illness control. Raising geese may be a fulfilling and sustainable agricultural activity if done with the right preparation and attention.

18

CHAPTER TWO

Choosing The Correct Breed For Farming Geese

Selecting the appropriate breed is one of the most important choices you will have to make when beginning a geese farming operation. The size, temperament, production, and environmental appropriateness of various breeds of geese vary substantially. This chapter will cover typical farming breeds of geese, breed selection considerations, and how temperament and production variations might impact your farming endeavors.

Popular Breeds Of Geese For Farming

A variety of breeds of geese are used in goose farming; each has special qualities that fit it for a certain purpose. While certain varieties are prized for their ability to lay eggs or for their aesthetic appeal, others are best suited for producing meat.

• Embden Geese

Because of their size, Embden geese are one of the most widely used breeds for industrial meat production. A popular option for farmers seeking high meat yields, Embden geese are distinguished by their white feathers and sturdy frame, with a maximum weight of 30 pounds (13.6 kg). They are perfect for novices since they grow fast and are reasonably straightforward to handle.

• Geese de Toulouse

The Toulouse goose, another heavyweight breed valued for its meat output, is French in origin. It is well-known for its capacity to acquire weight quickly and has a placid demeanor. Some types, like the Dewlap Toulouse, are very huge and are thus staples among farmers who seek a peaceful flock in addition to premium meat.

• Geese from Africa

African geese, despite their name, originated in Asia and are distinguished by the bump at the base of their beak. They usually weigh between 18 and 22 pounds (8-10 kg), which makes them smaller than Toulouse and Embden geese. African geese are a flexible option for farmers in numerous places due to their resilience and ability to adapt to varied conditions. Their ability to lay eggs makes them valuable as well.

• The pilgrim geese

Due to their auto-sexing characteristic, which allows males and females to be clearly identified from birth by color, pilgrim geese are unique. Males of this medium-sized breed are usually white, while females are usually grey. Because of their peaceful disposition, pilgrim geese are a great option for farmers who value handling simplicity in addition to a reasonable level of meat and egg output.

• Geese from China

Despite their tiny size, Chinese geese may lay up to 50–60 eggs annually, making them very prolific egg layers. They are well-known for being gregarious and vigilant, which makes them excellent property guards as they will honk loudly if they see someone breaking in. Because of their diminutive size, they are less often utilized to produce meat, but in certain markets, their eggs are highly sought after.

Things To Take Into Account While Selecting A Breed

Beyond personal taste, there are a number of considerations to take into account when choosing a breed of geese to make sure they complement your agricultural circumstances and objectives.

• The Reason for Farming

The first thing to think about is why you are raising geese. Do you want to raise geese mainly for meat, eggs, or as a hobby? Large breeds like Toulouse or Embden are great for producing meat. Breeds like Chinese or African geese might be more appropriate for you if your goal is egg production.

- **Harmony with the Climate**

Not every kind of goose is climate-appropriate. Certain species, such as African geese, are well-known for their ability to adapt to a wide range of environmental conditions, which makes them ideal for farmers in areas with variable temperatures. In contrast, because of their bulkier build and possible health risks associated with excess fat, Toulouse geese—especially the Dewlap variety—may need extra attention in colder climes.

- **Type of Land and Farm Space**

The best breed for your farm also depends on the size and kind of land you have available. Smaller breeds, like Chinese or Pilgrim geese, may survive in more cramped areas, while larger types, like Embden, need more freedom to go about and graze. Furthermore, varieties of geese that love swimming, like the African goose, may benefit from access to these natural features since geese prefer to live near water sources like ponds or streams.

• Needs for Maintenance and Food

The care and feeding needs of different breeds vary. For instance, although Chinese goose breeds are often lower maintenance, varieties like the Toulouse may need a more specialized diet to avoid excessive fat buildup. It's crucial to choose a breed that you can consistently offer the care and nourishment required for the best possible development and health.

• Demand in the Market

It's important to take into account the local market's need for certain breeds of geese or geese goods before choosing a breed. Goose eggs could be more in demand in certain markets than meat in others. Finding out what breeds the local's favor might help you choose the one that will provide the most financial return.

Variations In Productivity And Temperament

There may be significant breed-specific differences in temperament, so it's vital to take that into account, particularly if you're a novice or have kids or other farm animals to take into account.

• Aggressive vs. Docile Breeds

Certain kinds of geese, including Toulouse and Pilgrim geese, are inherently more gentle and manageable. These breeds are perfect for farms where regular human contact is required since they are less prone to display hostile behavior. However, other varieties, like the Chinese geese, are known to be more possessive and aggressive, which may make them harder to handle, particularly for new owners.

• **Watchful Conduct**

Although all geese are recognized to be good guardians, certain breeds are better than others. Due to their exceptional alertness and vocalization, Chinese and African geese are great at scaring off predators and intruders. But this behavior may also make them harder to control in a family agricultural setting.

• **Production of Meat vs. Eggs**

Depending on the breed's main purpose, productivity varies. Compared to bigger meat breeds like the Embden, Chinese geese lay a notably higher number of eggs due to their prolific egg-layering nature. Breeds such as the

African or Pilgrim geese provide an excellent balance between modest meat output and egg production, if you're interested in farming for both purposes (egg production and meat).

• Weight Gain and Growth Rate

Critical parameters to consider when evaluating productivity for meat farming are the geese's ultimate weight and development rate. Breeds with a reputation for quick development and high ultimate weights, like the Embden and Toulouse, are perfect for producers looking to maximize meat output quickly.

But in order for these breeds to reach their ideal weight, they may need more food and attention.

In conclusion, thorough consideration of a number of elements, such as your farm's aim, climate, available area, and market needs, is necessary when choosing the correct breed for goose farming.

By being aware of the temperamental and productive variations amongst popular breeds of geese, you can be sure that the geese you choose will fulfill your farming objectives and maintain a lucrative and manageable business.

CHAPTER THREE

Establishing The Farm

Perfect Area And Dwelling For Geese

Choosing the Proper Site

The health and production of your geese farm greatly depend on the site you choose. A large area is necessary for geese to graze, feed, and exercise. Since geese are excellent swimmers and like areas where they may bathe, the farm should ideally be located near freshwater sources, such as ponds or streams. In order to avoid waterlogging, which may cause health problems, the ground has to have good drainage. In order to keep water away from the geese's dwelling space and maintain it dry and pleasant, a gradual slope is advantageous.

Space Needs

Geese need plenty of room to move about. If geese are kept in pens or other enclosed spaces, it is generally recommended to provide them at least 10 to 20 square feet each. More room, however, is usually preferable since it fosters natural behaviors and lowers stress. Establish distinct areas for grazing, resting, and bathing if at all feasible. Geese will benefit from an open field that offers access to shady spots and cover from the rain and heat.

Design of Housing

Although they don't need fancy housing, geese should have a secure and useful space. It is sufficient to have a basic, well-ventilated shed or barn with strong walls and a roof. Make sure the housing has enough room for all the geese you want to keep, plus some extra for walking about. To avoid the accumulation of ammonia from waste, which may cause respiratory problems, proper ventilation is crucial. Positioning windows or vents to let in fresh air while minimizing drafts is recommended.

Constructing Adequate Pens And Shelters

Building of Shelters

When building geese shelters, think about using materials that are simple to maintain and long-lasting. A raised floor in a wooden or metal construction might assist keep the shelter dry and insect-free. To keep the geese safe from predators, the shelter needs a well-designed drainage system and a door that is secure. The shelter's inside should be maintained tidy and dry, and bedding made of hay or straw should be used to both absorb moisture and give comfort. Maintaining a healthy atmosphere and preventing manure accumulation need regular cleaning.

Pen Design

Geese should be able to wander about freely in their pen or enclosure while yet being protected from predators. To stop geese from fleeing or predators from entering, fencing has to be strong and tall enough. It is advised to have a fence that is at least 4 to 6 feet high. Furthermore, think about burying the fence's bottom a few inches below ground to prevent predators from tunneling under it. The enclosure should be large enough to avoid overcrowding, which may cause stress and health problems, and it should have access to water and shelter.

Comfort and Enrichment

Give geese in their enclosures enrichment to keep them happy and healthy. This may include places to forage, such as grass or clover patches, as well as places to rest or climb. Small ponds or troughs where they may wash and brush their feathers are a favorite feature for geese. The general health of the geese may be enhanced by routinely replacing the bedding and making sure the water supplies are pure and fresh.

Sustaining Safety And Cleanliness
Everyday Upkeep

Keeping the geese farm clean is essential to avoiding illness and guaranteeing the animals' well-being. Regular chores should include cleaning out dirty bedding, checking for wear and damage to the shelter and enclosures, and replacing food and water. Preventing health problems associated with contaminated food and water may be achieved by keeping the feeding facilities clean and clear of mold or rotting.

Management of Wastes

Sustaining a healthy ecosystem requires efficient waste management. Regular collection and appropriate disposal of manure are necessary. Composting is a great way to manage trash since it creates rich compost that can be used to fertilize gardens or crops, while also reducing the amount of garbage generated. In order to avoid contamination and pest attraction, make sure that composting sites are situated far from the geese's living quarters.

Biosecurity and Pest Management

Keeping a farm tidy and safe requires effective pest management. Check the pens and shelters on a regular basis for indications of pests like insects, rodents, or parasites. Pests may be avoided by putting into practice strategies including appropriate waste disposal, the use of pest deterrents, and upholding acceptable hygiene standards. To stop the spread of illness, biosecurity precautions including restricting access to the farm and cleaning boots and equipment are crucial.

Being Ready for Emergencies

Be ready for any emergency that may arise, such as severe weather or a disease epidemic. Make sure you have supplies, including feed, water, and medicines, on hand and have a strategy in place for moving the geese if required. In order to handle any new hazards or modifications to the agricultural environment, evaluate and update your emergency plan on a regular basis.

CHAPTER FOUR

Nutrition And Feeding

Nutrition and feeding are essential elements of a successful geese farming operation. Geese that are managed well will develop well, lay eggs effectively, and not get any diseases. Comprehending the dietary requirements, feed varieties, and feeding schedules will have a significant influence on your flock's output and welfare.

Geese's Nutritional Requirements

Similar to other fowl, geese have certain dietary needs that must be satisfied in order to maintain healthy and optimum development. They need a balanced diet that consists of a variety of fats, proteins, carbs, and vitamins and minerals.

Proteins are necessary for the formation of feathers, growth, and egg production. Goslings, or young geese, need more protein in their diet than adult geese. A high-protein diet aids in the quick development of feather and muscle growth. Fish meal, chicken byproducts, and soybean meal are common sources of protein.

Energy from carbohydrates is needed for everyday tasks. Grains are great sources of carbohydrates and should account for a large amount of a person's diet. Examples of grains include maize, wheat, and barley. The geese's metabolic functions are supported and their general health is preserved by these high-energy meals.

In addition to being essential for energy, fats also help absorb vitamins that are soluble in fat. However, in order to avoid obesity and associated health problems, the fat content should be reduced. In moderation, animal fats or vegetable oils may be a part of the diet.

Minerals and vitamins are essential for sustaining a number of body processes. Vitamins A, D, and E are crucial for healthy bones, immune system performance, and eyesight. The formation of bones and the generation of eggs depends on minerals like calcium and phosphorus. These vitamins and minerals are often included in balanced levels in well-formulated commercial feed, but you should also make sure geese have access to fresh water and grit.

In conclusion, high-quality proteins, carbs that are abundant in energy, the right fats, and vital vitamins and minerals are all part of a balanced diet for geese. To make sure they flourish, keep a close eye on their health and modify their food as necessary.

Varieties Of Feed (Supplements, Greens, And Grains)

Selecting the appropriate feed varieties is essential to fulfilling the varied dietary requirements of geese. A well-rounded diet may be achieved by combining grains, greens, and supplements since different feeds provide different nutrients.

For geese, grains are the main source of energy. Typical grains found in geese diets include oats, maize, wheat, and barley. Depending on the age and size of the geese, these grains may be given whole or processed into pellets or crumbles. Adult geese seem to do better on whole grains, whereas goslings do well with ground or crushed grains.

Minerals and vitamins are abundant in greens. The geese's food should consist of fresh pasture, clover, dandelion leaves, and other lush greens. In addition to offering vital minerals, spinach may assist add fiber, calcium, vitamins A and C, and other nutrients to the diet. Geese benefit from foraging possibilities because they may feed on native plants, which enhances their general health and well-being.

Supplements are used to improve diets or fill up particular nutritional deficiencies. These might be natural additions or commercial poultry nutrients. Typical add-ons consist of:

• Supplemental calcium: Required for laying geese to develop robust eggshells.

• Supplements containing protein, such as fish or soybean meal, to increase consumption of protein during times of high production or development.

• Supplements of vitamins and minerals: To make sure the geese get all the nutrition they need, particularly if their main meal isn't providing them with enough.

When using commercial supplements, it's crucial to adhere to the manufacturer's instructions. You should also get guidance from a veterinarian or a poultry nutritionist about the right kinds and dosages of supplements.

In conclusion, a varied diet that includes greens, grains, and supplements will help guarantee that geese get all the nutrients they need for good health and maximum output. A well-balanced mixture will promote their development, egg production, and general health.

Feeding Schedules For Various Age Groups

Feeding regimens should be adjusted based on the geese's age and growth stage in order to suit their unique nutritional needs. Goslings, developing juveniles, and adult geese have quite different demands.

Goslings (0–6 weeks): Due to their quick development, goslings need more protein. There should be a high-protein starting feed (18–20% protein). Small, frequent meals should be provided throughout the day to meet their high energy and development requirements. Always make sure kids have access to fresh, clean water since it is vital for healthy digestion and general well-being.

Growing Geese (6–16 weeks): Grower feed, which has a slightly lower protein concentration (14–16%), may be added to the diet as the geese become bigger since their protein demands decrease with growth. To guarantee a balanced diet during this phase, it's also advantageous to introduce a range of cereals and vegetables. Even if feeding frequency is lowered to twice or three times a day, nutrient-dense food is still necessary to promote ongoing development.

Adult Geese (16 weeks and older): A maintenance diet containing 12–14% protein is suitable for adult geese. They should eat a variety of cereals, vegetables, and sometimes supplements in their diet. Feeding may occur twice daily, with the goals of preserving body weight and promoting egg production. Females may need more calcium and protein throughout the breeding season in order to assist in the production of eggs.

Laying Geese: To maintain robust eggshells, it's critical to feed geese who are laying eggs a balanced diet that contains greater amounts of calcium (about 2-3% of the diet). It is advised to use a layer feed that contains 14–16% protein and additional calcium. The feeding plan is still in place, emphasizing access to clean water and regular feeding.

In conclusion, it is critical to the health and productivity of the geese that the feeding schedule and food be modified based on their age and stage of production. A healthy and productive flock may be maintained by routinely evaluating their condition and modifying their diet.

CHAPTER FIVE

Hatching And Breeding

Incubation: Natural Vs. Artificial

There are two main methods for incubating goose eggs: natural incubation and artificial incubation. Every approach has distinct advantages and disadvantages, and your breeding program's success might be greatly impacted by the strategy you choose.

Organic Fertilisation

In natural incubation, the eggs are incubated by a broody goose. This approach is often preferred since it is less complicated and requires less human involvement. It is a broody goose's natural urge to sit on her eggs and maintain their warmth, rotating them from time to time to make sure the heat is distributed evenly. For the eggs and goslings, this procedure might be less stressful since it is mostly self-sufficient.

You must make sure a goose has a cozy and safe place to nest if you want her to become a broody. When broody geese are given a peaceful, isolated area and are not disturbed, they often have more success. To prevent predators from interfering with the nesting process, it is important to keep an eye on the location. Naturally occurring incubation, however, might be unexpected since not all geese will exhibit broody behavior or keep it during the incubation period.

Synthetic Fertilisation

Contrarily, artificial incubation necessitates the use of an incubator to regulate the environmental factors required for egg development. Higher hatch rates may arise from the more accurate control this approach offers over temperature, humidity, and egg flipping. Artificial incubators are available in several shapes and sizes, ranging from little tabletop devices to more substantial cabinet-style apparatuses.

For optimal gosling growth in an incubator, it is essential to maintain the proper temperature (typically about 99.5°F or 37.5°C) and humidity level (usually between 55 and 60%). To guarantee equal growth, eggs should be flipped frequently—typically at least three times each day. The procedure may be made simpler by the mechanical rotation mechanisms found in many contemporary incubators.

Selecting Between Incubation Methods: Natural And Artificial

The choice between natural and artificial incubation is based on a number of variables, such as your preferences, your geese experience, and the resources you have at your disposal. Although natural incubation may be less predictable, it may be more economical and need less technology. Although artificial incubation requires more financial outlay and careful equipment maintenance, it offers better control and consistency.

In the end, both approaches are viable if they are handled correctly. Some breeders combine the two procedures, using natural incubation when practical and artificial incubation when necessary to increase hatch rates or handle a higher volume of eggs.

Handling Breeding Pairs

An effective management of breeding pairs is necessary for a healthy and successful geese farm. The procedure includes choosing compatible partners, making sure housing is adequate, and keeping oneself healthy to promote successful reproduction.

Choosing Breeding Pairs

Choosing geese with desired features is the first stage in managing breeding pairs. Seek for robust, genetically sound, and aesthetically pleasing birds. Geese should be at the proper age since distinct breeding behaviors might be seen in young and elderly birds. Geese usually begin to breed around the age of two, but if they are in excellent condition, older birds may still be prolific.

It's crucial to take the geese's disposition into account. A geese's capacity to establish a successful breeding pair may be hampered by their tendency towards aggression or dominance. It is possible to predict if prospective mates will get along and work together throughout the breeding season by keeping an eye on their behavior.

Environment and Housing

It is essential to provide breeding partners with an appropriate habitat. Geese need a large, safe space where they may live in comfort. The place used for nesting has to be protected from the weather and kept unaltered. An atmosphere that is conducive to laying and incubation may be created with the use of a clean, dry nesting box and enough bedding.

Make sure the breeding couples have access to a balanced meal and clean water as well. In order to promote general health and fertility, effective reproduction, and healthy progeny, a proper diet is essential.

Welfare And Health

Maintaining good health is essential for breeding geese. Make sure they are clear of parasites and illnesses since these might have a detrimental effect on the effectiveness of breeding. A veterinary professional's suggestions should be followed while administering vaccinations and preventative care.

Keep a careful eye out for any indications of stress or hostility in the geese throughout the mating season. To keep the atmosphere peaceful, take quick action to resolve any problems. Infections may be avoided and a healthy breeding process can be encouraged by routinely cleaning the nesting space and adding new bedding.

Taking Care of Goslings and Eggs

To guarantee healthy goslings, appropriate care must be given once the eggs are deposited. During this phase, eggs must be handled carefully, ideal circumstances must be maintained, and the freshly born goslings must get the necessary care.

Taking Care of and Keeping Eggs

To prevent harm, eggs should be carefully gathered after they are deposited. Eggs should be kept dry and cold until they are ready to be incubated. The protective covering on the eggs may be removed by washing, which raises the possibility of infection. If cleaning is necessary, carefully wipe away any dirt with a dry cloth.

Care for Incubation

Preserving the ideal circumstances is essential whether incubation is done naturally or artificially. Check the humidity and temperature in your nesting space or incubator on a regular basis.

Keep an eye out for any indications of growth in the eggs and flip them as needed. Periodically candling the eggs can aid in monitoring their development and spotting any problems early on.

Taking Care of Goslings

To guarantee their survival and well-being, the goslings need to be cared for as soon as they hatch. The goslings need a warm, dry environment since they are sensitive to temperature fluctuations. A brooder with a heating element may assist in maintaining the right temperature.

Goslings need access to pure water and nutrient-rich, age-appropriate food. Keep a careful eye on their growth and development and make any necessary adjustments to their habitat and nutrition.

To assist goslings feel safe and less stressed, make sure they have companions. Social engagement is also crucial.

Reduce the brooder's temperature gradually as the goslings develop, and when they're ready, take them outside. Once they mature from goslings to adult geese, keep an eye on their health and modify their treatment as needed.

CHAPTER SIX

Management Of Health And Disease

Typical Health Problems With Geese

Similar to other livestock, geese are prone to various health problems that may impact their overall health and efficiency.

Comprehending these prevalent health issues is essential for efficient treatment and avoidance.

1. infected respiratory systems:

Respiratory diseases are common in geese, especially in moist or poorly ventilated habitats. Aspergillosis and avian influenza are common respiratory illnesses.

Breathing difficulties, nasal discharge, and coughing are possible symptoms. Respiratory issues may be prevented by keeping their home clean and well-ventilated.

2. Invertebrates:

In geese, both internal and external parasites may seriously compromise their health. Skin discomfort and the loss of feathers may be caused by external parasites like lice and mites. Worms and other internal parasites may lead to diarrhea, weight loss, and overall lethargic behavior. To maintain the health of geese, regular deworming and treatment for external parasites is important.

3. Foot Issues:

Geese need healthy feet since they spend a lot of time walking and standing. Bumblefoot is a common foot ailment that results in painful, swollen feet due to bacterial infections. Foot problems may be made worse by unhygienic and moist flooring in poor living circumstances. Foot disorders may be avoided by giving clean, dry bedding and doing routine foot examinations.

4. gastrointestinal disorders

Digestive diseases including gizzard erosion and enteritis may affect geese. Lethargic behavior, low appetite, and diarrhea are possible symptoms. The two most important preventative actions are to provide access to clean water and to a balanced diet. Steer clear of abrupt dietary changes since they may cause digestive system disturbances.

5. Problems with Egg-Laying:

Egg-laying problems may be a difficulty for laying geese. It is possible to have issues like egg binding, in which the egg becomes lodged in the reproductive canal. Egg-laying problems may be avoided by making sure geese have a healthy diet that includes enough calcium and by creating a stress-free habitat.

Health Care Preventive (Vaccines, Hygiene)

Maintaining the health of geese and averting illness outbreaks need effective preventive healthcare. Ensuring the flock's long-term health may be achieved by putting a thorough health management strategy into action.

1. Immunisations:

One of the most important aspects of preventive healthcare is vaccination. Vaccines may shield geese against a number of illnesses, such as salmonella, Newcastle disease, and avian influenza. To create a vaccination program that is specific to the requirements of your flock, speak with a veterinarian. Frequent vaccination lowers the chance of disease spread and helps stop epidemics.

2. hygienic habits

Disease prevention in geese is contingent upon maintaining optimal cleanliness. To lower the risk of illness, clean and sanitize housing, feed containers, and water supplies on a regular basis. Make sure all equipment is well cleansed and use non-toxic disinfectants that won't harm geese. To reduce the chance of disease transmission, routinely remove manure and other waste from the living environment.

3. Biosecurity Procedures:

By putting biosecurity measures in place, illnesses may be stopped from entering the environment and spreading. Minimize the amount of time your geese interact with other birds, and refrain from adding additional birds to your flock without following the correct quarantine protocols.

Make sure guests observe biosecurity procedures, such as cleaning shoes and gear before approaching the geese's territory.

4. Diet and Nutrition:

Giving geese a varied diet is essential to their well-being. Give your geese a nutritionally balanced meal that is tailored to their individual requirements and has all the vitamins and minerals they need. Make sure the geese are always able to obtain fresh, clean water. Steer clear of abrupt dietary changes as they may result in stomach troubles and other health concerns.

5. Frequent Health Examinations:

Check your geese's health on a regular basis to see any symptoms of disease early. Keep an eye on their general health, appetite, and behavior. Check for anomalies on a regular basis in their feet, droppings, and feathers. The health and welfare of your flock may be guaranteed by identifying and treating health concerns early on.

Identifying Symptoms Of Illness

Understanding the symptoms of sickness in geese is essential for prompt intervention and care. Knowing the typical signs can enable you to respond appropriately and quickly to health issues.

1. Modifications in Conduct:

A shift in behavior is among the first indicators of disease in geese

. A sick goose may lose interest in food and drink, become sluggish, or separate from the flock. Keep an eye out for any deviations from their typical behavior patterns and take prompt action if you see any odd signs.

2. Acute Airways Complaints:

In geese, respiratory problems may point to an infection or other medical condition. Keep an eye out for symptoms including breathing difficulties, nasal discharge, coughing, or sneezing.

There may be edema in the face or around the eyes along with these symptoms. Treating respiratory problems as soon as possible is crucial to keeping them from becoming worse.

3. Issues with the Digestive System:

Changes in feces, such as blood in the stool or diarrhea, might be signs of digestive issues. A reduction in appetite or indications of stomach pain may potentially be markers of illness in geese. Keep a close eye on their droppings and get advice from a veterinarian if you see any noticeable changes.

4. Skin Conditions and Feathers:

Check your geese's feathers and skin on a regular basis. Feathers falling off, strange skin sores, or weird growths are some indicators of disease. External parasites like lice or mites may damage feathers and irritate the skin. Any anomalies in the feathers or skin should be addressed right as to avoid further problems.

5. Loss of Weight and Weakness:

Geese may have a health problem if they are becoming thinner or exhibiting indications of frailty. Keep an eye on their physical state and make sure they are keeping a healthy weight. Weakness and weight loss might be signs of a number of health issues, such as parasites, illnesses, or malnutrition.

CHAPTER SEVEN

Controlling Geese Conduct

Recognizing The Social Structures Of Geese

Similar to several other bird species, geese have intricate social systems that are crucial to their behavior and overall welfare. Anyone who farms geese must comprehend these systems in order to control their behavior and maintain a peaceful atmosphere.

1. The Value of Structure

An essential component of geese social interactions is the establishment of a pecking order within their flocks. Access to resources like food, water, and nesting locations is determined by this hierarchy. The dominant geese, often known as "alpha" geese, typically lead the flock and get first dibs on these resources. Comprehending this hierarchy may facilitate the resolution of disputes and provide equitable access to resources for all geese.

2. Family Units and Pair Bonding

Geese are recognized for their ability to create enduring pair relationships. These ties are crucial to their social structure because they often result in the flock's family units being formed. These families have the power to shape behavior and group dynamics. A mated couple, for example, will be more protective of their young and may act more aggressively towards strangers or intruders.

3. Social Signals and Communication

Geese interact with one another using a range of vocalizations and body signals. It is essential to comprehend these signals in order to explain their behavior. For instance, calling and honking are often used to coordinate movement or warn the flock of possible dangers. You can better handle and attend to your geese's requirements if you are familiar with these indications.

4. Including the New Geese

Care must be taken when bringing in young geese to an established flock to avoid upsetting the social order. It's crucial to progressively add more geese and see how they interact with the current flock. This promotes a less hostile environment and facilitates a more seamless integration process.

5. Social Behaviour and the Environment

Geese's social behavior may be greatly influenced by their living conditions. The way geese interact with one another may be influenced by elements including area, cover, and the existence of water sources. Reducing behavioral problems and preserving a stable social structure may be achieved by creating an atmosphere that suits their demands.

Managing Forceful Or Defence Behaviour

Farmers may find it difficult to deal with aggressive or protective behavior in geese, yet doing so is essential to preserving a secure and fruitful environment.

1. Recognising Aggressive Conduct

Geese may exhibit aggression in a number of ways, including pursuing, biting, and loud vocalizations. This conduct is often a reaction to things that they see as challenges or dangers to their social standing. Effective management of violent behavior begins with identifying its causes.

2. Taking Care of Aggressive Conduct

As soon as violent behavior is seen, it must be addressed right away. To avoid such mishaps, this may include removing the belligerent geese from the flock. Over time, gradual reintroduction along with supervision and monitoring may aid in the reduction of hostility.

3. Behaviour Protective of Offspring

Geese are renowned for having strong protective tendencies, particularly towards their young. Despite being normal, this behavior sometimes results in hostile confrontations with people or other animals. Managing protective behavior may be aided by making sure the locations used for nesting and raising are safe and have enough room.

4. Conditioning and Training

When it comes to controlling violent behavior, training and conditioning may be useful techniques. More desired behaviors may be shaped with the use of positive reinforcement strategies, such as rewarding calm behavior with treats or other incentives. Patience and consistency are essential throughout this procedure.

5. Getting Expert Guidance

When handling gets difficult or there is extreme hostility, consulting a professional or an experienced geese farmer might be helpful. They may provide perceptions and methods for controlling hostile conduct and enhancing flock management in general.

Educating And Developing Your Geese

Geese may be trained and tamed to improve their manageability and increase the enjoyment of interactions with them. Even though geese are typically bright and able to learn, training takes time and dedication.

1. Establishing Trust

Establishing trust is the first stage in teaching geese. Creating a strong relationship with your geese may be facilitated by consistent, pleasant encounters including feeding and gentle touching. Building trust with your geese is crucial to their training success and will make them more open to picking up new skills.

2. Basic Instructional Guidelines and Methods

Managing geese may be made simpler by teaching them to obey simple orders. Begin with basic directives like "come" or "stay," then use incentives like praise or rewards to reinforce desirable behavior. Reward and command consistency is essential to successful training.

3. Getting Geese Used to Each Other

A crucial part of teaching geese is introducing them to people and other animals. By gradually exposing them to a variety of stimuli, fear and anxiety may be lessened and the person becomes more situationally adaptive. During socialization, positive reinforcement might help them become even more confident and behave better.

4. Handling Obstacles

There may be obstacles in the training process, such as a phase of fear or noncompliance. It's critical to handle these setbacks with tolerance and compassion. To overcome these obstacles, reassess your workout regimen and make any necessary modifications.

5. Sustaining Constant Training

Training is a constant effort that is part of keeping your geese under control. Consistent reinforcement of behaviors and directives aids in sustaining the gains obtained during the first training. Your geese will continue to be attentive and well-behaved if you provide them with plenty of attention and positive reinforcement.

You may foster a more peaceful atmosphere for your flock and have a more fulfilling experience with geese farming by learning about the behavior of geese and putting good management techniques into practice.

CHAPTER EIGHT

Managing Grazing And Water In Geese Farming

Water Is Essential For Geese

Water plays a major role in the everyday lives of geese, affecting their general well-being, behavior, and health. Geese, in contrast to some other livestock, need a lot of water for both drinking and engaging in their natural activities, such as swimming and foraging. Maintaining the health of your flock requires making sure there is a steady supply of clean water.

Needs for Drinking Water

Due to their size and metabolic requirements, geese need a significant amount of drinking water. They must always have access to clean, fresh water. Inadequate water consumption might cause general health issues, decreased egg production, and dehydration. Geese typically drink one to two liters of water a day, depending on their size and the weather outside. Keeping water available to them all the time keeps their digestive systems healthy and helps them remain hydrated.

Aspects of Behaviour

The natural behaviors of geese are significantly influenced by water. They are known to partake in water-related behaviors like bathing and dabbing, which contribute to the cleanliness and parasite-free state of their feathers. Not only are these enjoyable, but they are also necessary to keep their feathers healthy and stop dangerous bacteria from growing.

Considering Health

If their water supplies are not well maintained, geese are vulnerable to parasites and illnesses spread by the water. To avoid pollution, water troughs and ponds need to be cleaned and maintained on a regular basis. Furthermore, it is crucial to guarantee that the water is devoid of contaminants and dangerous substances in order to prevent health problems.

Water Availability

It's crucial to offer geese readily available water sources during farming. This might apply to troughs, tanks, or organic pools of water. These water sources should be positioned strategically so that no geese will be unable to reach them due to competition or congestion. It may be necessary to provide many water stations in bigger flocks in order to adequately accommodate every bird.

Having swimming areas or ponds

Ponds and other sources of water are vital to geese everyday existence, thus they are naturally drawn to them and utilize them. In addition to promoting their natural behaviors, having ponds or swimming places on your farm improves their enjoyment and health.

Water Body Types

Depending on the area and resources you have available, you may provide a variety of water bodies. Small ponds, man-made lakes, or even enormous water-filled tanks are among the options. Every kind has unique advantages and things to think about.

• Small Ponds: These may be made by excavating a specific area or by using natural landscapes, and they are great for smaller flocks. They provide geese with a more natural setting and support their innate tendencies.

• Artificial Lakes: Bigger flocks may make use of artificial lakes, which are larger bodies of water with plenty of area for swimming and foraging. Although they need more upkeep, they can house more geese.

• Big Tanks: Big water tanks may provide good swimming areas in places where it is not practical to build natural ponds. To stop the growth of algae and other pollutants, these tanks should be cleaned on a regular basis.

Upkeep and Security

Regular cleaning and observation are necessary for the upkeep of swimming pools and ponds. It's common knowledge that geese may churn up silt, which, if left unchecked, can result in murky water. Water may be kept clean by routinely looking for and eliminating debris, making sure tanks are properly filtered, and so on.

Another issue is safety, particularly in man-made ponds. Make sure there are no risks or sharp edges that might hurt the geese. Creating accessible access points or moderate slopes may make it safer for geese to enter and depart the water.

Advantages For Geese Health

Geese benefit in several ways from having access to ponds or swimming areas:

• Feather Care: Swimming assists geese in maintaining the cleanliness and health of their feathers. Feathers that are clean are necessary for buoyancy and insulation.

• Parasite Control: Regular swimmers have fewer external parasites since the water helps wash and remove parasites from their feathers.

• activity and Enrichment: Swimming reduces boredom and promotes natural behaviors by offering both mental and physical activity.

Keeping Pastures and Grazing Geese

Grazing is essential to goose farming since it keeps your flock healthy and helps to offer a natural diet. Good pasture management helps manage the land responsibly and guarantees geese a steady supply of nutrient-rich fodder.

Grazing Advantages

Since they are grazers by nature, geese thrive on a range of forages, including legumes and grasses. Grazing improves their general nutrition and keeps their digestive systems in good working order. Moreover, grazing gives geese an opportunity for exercise and lets them exhibit their natural behaviors.

Pasture Administration

Several crucial measures are necessary for effective pasture management:

• Rotational Grazing: Putting in place a rotational grazing system may assist in keeping pastures healthy and preventing overgrazing. To help the grasses recover, this entails splitting the pasture into portions and alternating the geese between them.

• Pasture Fertility: Ensuring the development of nutrient-dense fodder requires routinely evaluating and enhancing pasture fertility. To keep the soil healthy, this might include testing the soil and adding the right amount of lime or fertilizers.

• Weed and Pest Control: Keeping pastures free of weeds and pests is essential. Avoid hurting the geese by managing invasive plants and pests using organic approaches or carefully chosen treatments.

Sustaining The Quality Of Pasture

Maintaining pastures in excellent shape requires:

• Frequent Mowing: Mowing lawns on a regular basis encourages healthy development and helps maintain grass at the ideal height. Moreover, it prevents weeds from taking over.

• Reseeding: Reseeding pastures on a regular basis may assist in maintaining a steady supply of nutrient-rich grasses and enhance the quality of the fodder.

• Water Access: To sustain grazing geese, make sure pastures have sufficient access to water sources. They must be well-hydrated for both general health and productivity.

Ecological Methods

Using sustainable pasture management techniques is crucial for the environment's and your geese's long-term well-being. This comprises:

• Reducing Soil Erosion: Reducing soil erosion and degradation may be achieved by limiting overgrazing and preserving ground cover.

• Promoting Biodiversity: A variety of plant species in pastures helps feed geese and enhance the condition of the soil.

• Waste Management: Maintaining pasture health and minimizing environmental effects may be achieved via the appropriate management of manure and waste products.

CHAPTER NINE

Gathering Products From Goose

Gathering And Utilising Eggs

A crucial component of raising geese is egg harvesting, which calls for meticulous attention to detail and consistency. Because they produce a lot of eggs, geese may be valued commodities in specialty and local markets. Understanding the reproductive cycle and making sure the geese are in good condition to produce high-quality eggs are the first steps in the procedure.

1. When and How to Gather

Every female goose lays anywhere from 20 to 50 eggs during a single nesting season, which usually occurs in the spring. Typically, the eggs are placed in nests, often in isolated locations. Checking the nests on a regular basis is essential. Gathering eggs is best done early in the morning before the geese go back to their nests. By using this technique, the chance of contamination and harm is reduced.

2. Managing and Keeping in Stock

Eggs should be handled carefully after they are gathered to prevent breaking. To get rid of any dirt or debris, gently wipe them down; however, don't wash them unless absolutely essential, since this might damage the protective layer. The eggs should be kept dry and cold. To preserve their quality, they should be stored at a temperature of 45–55°F (7–13°C), in a humidity-controlled atmosphere.

3. Applications and Future Market Opportunities

Gourmet chefs and food fans are drawn to the different flavors and bigger, richer size of goose eggs compared to chicken eggs. They may be used in savory meals, baking, and custard recipes, among other culinary endeavors.

Furthermore, goose eggs are often promoted as a specialty item that may fetch premium costs in gourmet shops and farmers' markets.

Gathering Feathers For Down

Another crucial component of raising geese is collecting their feathers, particularly for those who manufacture goods including down and feathers. The soft underlayer of feathers called down is highly valued for its insulating qualities and is utilized in many goods like as outdoor clothing and bedding.

1. Recognising Down from Feathers

Understanding the difference between down and feathers is crucial. Soft, fluffy clusters that trap air and act as insulation make up down. In contrast, feathers are more stiff and feature a central quill. Both are valuable commercially, although down is often more in demand due to its better insulating qualities.

2. Techniques for Harvesting

There are two main methods for gathering down: plucking and molting. The process of plucking, which is done by hand, generally occurs after the bird has been killed. Although labor-intensive,

this procedure produces down of excellent quality. Gathering down from geese during their regular feather-shedding cycle, known as molting, is the second way. This is a less intrusive procedure that you may do again every year.

3. Procedures and Inspection

Feathers and down must be cleaned and separated after collecting. Washing and drying are steps in the cleaning process that get rid of any contaminants. Based on the kind and grade of the down or feathers, sorting is done. Superior grade down exhibits fluff and high fill power, signifying its capacity to act as an insulator. When downsized appropriately, it maintains its inherent qualities and may be used to produce high-end goods.

4. Getting Ready for the Market

Goosedown is in high demand in the outdoor gear and bedding sectors, creating a competitive industry. High-quality down may be sold directly to customers or to manufacturers. To preserve openness and build confidence with purchasers, it's critical to have thorough documentation of the procurement and processing of down. To appeal to ethical and environmentally sensitive customers, you should also think about promoting the ethical and environmental components of your harvesting methods.

Production Of Meat And Market Readying

An important aspect of the agricultural business is the production of meat from geese, which has to be managed carefully to guarantee premium meat for the market. Since this is when they reach their ideal weight, geese are reared mostly for their meat in the autumn and winter.

1. Growth and Feeding

The food and housing of geese have a significant impact on the meat's quality. For optimum development, geese should be given a well-balanced diet high in grains and protein. Care should be used while adjusting the feeding schedule to prevent underfeeding, which may cause underweight birds, or overfeeding, which can result in fatty meat.

2. Processing and Slaughtering

Geese should be slaughtered humanely and in compliance with local laws. To reduce stress, the birds are first stunned, then they are killed and their feathers removed. It is essential to handle the meat carefully at this phase in order to preserve its quality. The birds are processed, decapitated, and plucked after being killed. It's important to chill meat fast to keep it fresh and free of spoilage.

3. How to Package and Preserve

Goose meat must be carefully wrapped and stored for market processing. To avoid contamination and freezer burn, the meat must be vacuum-sealed or wrapped in moisture-resistant packaging.

To maintain its quality, the meat has to be frozen or refrigerated as soon as possible. A market appeal may be increased by clearly labeling products with the weight, processing date, and any applicable certifications.

4. Sales and Marketing

Because of its rich flavor and soft texture, goose meat is a specialty product that is often sought. It may be sold to upscale dining establishments, specialty meat stores, and to customers directly via the Internet or farmers' market channels. Emphasizing the meat's quality, the moral elements of your agricultural methods, and any special features might draw in customers. Developing connections with merchants and chefs may also lead to new business prospects.

CHAPTER FEN

Selling And Promoting Geese Goods

Locating Regional Markets For Geese Goods

Recognising Local Need

Finding local markets is essential to selling your goods successfully. Start by finding out how much local demand there is for your goods. To determine interest,

go to local fairs, farmers' markets, and community activities. To learn about the demands and preferences of the local chefs, restaurant owners, and specialty food shops, speak with them. Participate in neighborhood discussion boards and social media communities centered on regional food and farming to get information.

Investigating Community Events and Farmers' Markets

Farmers' markets are great places to offer things made from geese. Customers are interested in fine culinary goods supplied locally frequently in these marketplaces. Create a booth where you may sell your geese goods, such as meat, feathers, and fresh eggs. Make the most of this chance to engage with clients, get their opinions, and expand your clientele. Take part in regional fairs and events as well to spread the word about your farm to a wider audience.

collaborating with nearby eateries and specialty shops

Building ties with neighborhood eateries and specialty food shops may be very advantageous. Speak with eateries that specialize in specialty food made using animal products or farm-to-table eating. Provide samples and talk about how your goods may go well with their cuisine. In a similar vein, focus on specialty food shops that carry high-quality, regionally produced goods. Forming these alliances may increase the exposure of your farm and provide a consistent source of income.

Making Use of Online Resources

Leveraging online platforms is crucial for increasing your market reach in the current digital world. Establish a website or an e-commerce platform to exhibit your geese merchandise.

Make use of social media channels to interact with prospective consumers, communicate farm updates, and advertise your goods. Social media sites like Facebook and Instagram are great for visual marketing and can foster a brand community.

Ways To Price Meat, Eggs, And Feathers

Calculating Production Cost

The first step in accurate pricing is knowing your manufacturing costs. Compute the price of labor, feed, medical care, and any other costs associated with rearing geese. To achieve profitability, include these charges in your pricing-determining process. Maintain thorough records of every expenditure to aid in financial planning and price changes.

Competitive Pricing and Market Research

To find out how much comparable items in your region are selling for, do market research. Examine the costs of rivals for the meat, eggs, and feathers of geese. Make sure your costs are reasonable and accurately convey the superiority and distinctiveness of your offerings. Different client groups may be drawn in by providing a variety of price alternatives, such as bulk discounts or seasonal specials.

Pricing Based on Value

Think about value-based pricing, where the price is determined not only by manufacturing costs but also by the perceived worth of your items. Draw attention to certain features of your geese goods that might make them more expensive, such as heritage breeds or organic feed. To support your premium pricing, inform customers about the advantages and exceptional quality of your products.

Putting Flexible Pricing Strategies Into Practice

Pricing flexibility may assist in meeting a range of consumer demands and market circumstances. Use techniques like initial offers, package discounts, or subscription plans for loyal clients. To stay competitive and increase income, adjust pricing in response to changes in the market, seasonal variations in demand, and other variables.

Developing Your Farm's Brand
Formulating an Identifiable Brand Image

Creating a distinctive brand identity is crucial to setting your farm apart from rivals. Begin by coming up with a catchy farm name and logo that captures the spirit of your offerings and your ideals. Think about components that will appeal to your target demographic, such as color palettes, font, and artwork. Your brand should exude authenticity, dependability, and excellence.

How to Write an Engaging Brand Story

Developing a strong brand narrative helps in creating emotional bonds with consumers. Tell us about your farm's history, including how you got started raising geese, your dedication to excellence, and the special features of your goods. Tell this tale and establish a personal connection with your clients by using your website, social media accounts, and promotional materials.

Formulating a Marketing Plan

The foundation for creating and sustaining your brand is an efficient marketing plan. To market your company, use a variety of platforms, including influencer collaborations, social media, and local advertising. Provide informative material that engages readers, gives agricultural updates, and emphasizes the advantages of your goods. Maintaining a consistent brand across all platforms helps in strengthening the identification of your farm.

Developing Relationships with Customers

Developing trusting bonds with clients promotes recurring business and loyalty. Provide outstanding customer service, answer questions right away, and interact with clients via surveys and feedback. To reward loyal consumers and motivate them to tell others about your farm, put in place loyalty programs or referral bonuses.

Increasing the Visibility of Your Brand

Take into consideration working with other nearby companies or taking part in cooperative marketing initiatives to increase the visibility of your brand. Look into possibilities for co-branding or cross-promotional campaigns with related goods or services. In order to promote your goods and build relationships with possible partners and clients, you should also think about going to trade exhibitions or industry events.

CHAPTER ELEVEN

Growing Your Farm Of Geese

Growing Your Crowd

A crucial first step in growing your goose farm is increasing your flock, which calls for meticulous preparation and implementation to guarantee healthy, sustained expansion. Expanding entails not just boosting the quantity of geese but also modifying a variety of farm management practices.

Evaluating Your Available Resources

Consider if your present resources can sustain the addition of more geese before growing your flock. This involves determining how much room your grazing spaces, housing, and feed supply can accommodate. Make sure your facilities are up to date and adequate to hold additional geese without endangering their welfare. Think about your pastures' quality and the accessibility of water supplies as well.

Choosing the Correct Breed

Selecting the appropriate breed is crucial for growth. Choose a breed that fits your farm's objectives and surrounding surroundings since different breeds have different demands and traits. For example, you may choose to expand meat production by using breeds that are recognized for their quick development and high productivity. On the other hand, breeds with more egg-laying capacity would be better suited for egg production.

Breeding as well as Buying

Decide whether you will extend your flock by breeding or by acquiring additional geese. Breeding may be a cost-effective strategy, but it needs careful management to guarantee the genetic variety and health of the progeny. Purchasing new geese entails choosing reputable breeders or hatcheries to guarantee you obtain healthy and high-quality birds. To stop the spread of illness, always isolate new arrivals before integrating them with your current flock.

Adapting Facilities and Infrastructure

Your infrastructure and amenities will need to be modified as your flock grows. This entails increasing grazing grounds, renovating existing housing units, and making sure your water and feed infrastructure can manage the extra demand. To keep your geese in a healthy environment, housing units must have enough ventilation and heating.

Monitoring and Adjusting

After extending your flock, check the new recruits regularly for symptoms of stress or health concerns. Check their integration with the current flock on a regular basis and make the required modifications to avoid conflicts.

Keep track of production measures, like as egg output or weight increase, to assess the efficacy of your expansion and make data-driven choices for continued development.

Hiring Farm Help
Determine Staffing Requirements

As your geese farm expands, you will need to hire more staff in order to properly handle the growing task. To begin, list the precise duties and tasks that will need additional help. This might include maintaining records, checking health, cleaning, and feeding on a regular basis.

To guarantee efficient operations, clearly outline each position's duties and responsibilities.

Hiring and Instruction

Hiring competent and trustworthy agricultural labor is essential to preserving the effectiveness and caliber of your business. Think about publishing job postings on internet job boards, community boards, or regional agricultural forums. Seek applicants who have knowledge of animal husbandry or who have a desire to learn throughout the employment process. Give new recruits extensive training to make sure they can handle their jobs well and comprehend agricultural processes.

Overseeing Employees

Creating a healthy work atmosphere and having clear communication channels are essential to the effective management of agricultural personnel. Establish a method for allocating work and scheduling shifts to make sure that all duties are fulfilled.

Evaluate employee performance on a regular basis and provide constructive criticism to promote ongoing development. Retaining valuable personnel may also be aided by providing incentives or rewards.

Security and Adherence

Make certain that every employee understands and abides by the safety and compliance guidelines pertaining to the farming of geese. This includes using safety gear, managing birds properly, and following biosecurity protocols. Organize frequent safety training sessions and keep up-to-date documentation of adherence to regional laws and guidelines.

Assigning Accountabilities

As your farm grows, assigning tasks to others becomes more crucial. Assign distinct responsibilities to various team members according to their abilities and capabilities. For instance, one person may be in charge of health monitoring while another would be in charge of feed management. Delegating well facilitates operations and frees you up to concentrate on the important elements of the farm expansion strategy.

Taking Care Of Growing Sales And Production

Simplifying Processes

To effectively handle the increasing volume, you must streamline your processes in order to manage increased production and sales. Establish procedures and methods to monitor sales, manufacturing rates, and inventories. Adopting management systems or software solutions that can automate processes and provide real-time data on farm performance is something to think about.

Improving Sales and Marketing Approaches

In order to take advantage of higher output, improve your sales and marketing plans. To reach prospective clients, create a thorough marketing strategy that uses both physical and online platforms. To sell your geese goods, use farm tours, local markets, and social media. Developing trusting connections with nearby stores, eateries, and companies may also aid in growing your clientele.

Assessing Monetary Performance

Assess your financial performance on a regular basis to make sure that higher output is generating profits. Keep an eye on your costs, income, and profit margins to spot areas where you may cut costs or make investments. For assistance in creating successful budgeting and cash flow management techniques, think about collaborating with an accountant or financial adviser.

Customer feedback and quality assurance

Upscaling requires you to maintain high standards of quality. Put quality control procedures in place to guarantee that your geese and their output live up to consumer and industry requirements. Get client feedback so you can learn about their preferences and resolve any issues. Constantly improving your business in response to client feedback may boost your standing and increase revenue.

Organizing for Potential Growth

As you oversee more output and sales, make plans for future expansion by establishing long-term objectives and creating plans to meet them. Think about future technological, facility, or product expansions. To adjust and maintain your competitive edge in the market, keep up to date on developments and trends in the business.

Conclusion
Considering Geese Farming

It's important to take stock of our trip as we near the conclusion of our investigation into novice geese farming. Establishing a geese farm entails more than simply rearing birds; it's about committing to a way of life marked by commitment, endurance, and a profound love of the natural world. Anybody wishing to enter this fulfilling sector should find a thorough foundation from the ideas and techniques covered in this book.

Dedicated to Sustainable Methods

The significance of sustainable practices is an important lesson to learn from this book. When carried out properly, goose farming has the potential to be a very sustainable activity. By putting the health and welfare of the geese first, using resources wisely, and using environmentally responsible techniques, you can establish a farm that serves your objectives and benefits the environment at the same time. This strategy not only protects your farm's long-term profitability but also adheres to more general environmental principles.

Creating a Community

The importance of community is emphasized in this book as another important factor. Participating in forums, joining pertinent organizations, and establishing connections with other goose farmers may all provide insightful and helpful information. These relationships may be crucial for conquering obstacles, celebrating victories, and never stopping learning. Creating a network of mentors and peers is one of the most satisfying things about beginning your adventure into goose farming.

Taking the Learning Curve in Hand

The sector of goose farming is dynamic and offers opportunities for lifelong growth. You can run across unanticipated possibilities and obstacles while putting the techniques and tactics covered into practice. Accept these encounters as a necessary part of the learning process. Keep up with changes in poultry care, industry trends, and developing best practices. Long-term success will depend greatly on your ability to change and develop.

Honouring Achievements

Celebrate all of your accomplishments, no matter how little. Every accomplishment—a robust flock, a bountiful crop, or a delighted client—is evidence of your diligence and commitment. Acknowledge these successes and let them inspire you to keep aiming higher.

In summary, learning to raise geese is an exciting and rewarding experience. Equipped with the information and abilities acquired from this book, you're ready to take on this thrilling endeavor. Recall that enthusiasm, tenacity, and ongoing education are the keys to success in geese farming. I hope your endeavor to raise geese will be rewarding and successful.

Learning To Farm Geese

Overview of Duck Farming

For novices, goose farming is a special chance to explore the world of poultry with an emphasis on a bird that is renowned for its appeal and adaptability. This section attempts to provide a basic overview of goose farming, emphasizing the advantages, difficulties, and prerequisite information needed to get started.

Knowing the Fundamentals

Understanding the fundamentals of poultry farming is essential to starting a profitable goose farming operation. Compared to other fowl, geese have unique demands and behaviors as a species. Ensuring their well-being and productivity requires an understanding of these essentials, which include their food demands, housing requirements, and social behaviors.

Selecting the Appropriate Breed

One of the first and most crucial choices in goose farming is choosing the right breed. The traits of different breeds of geese vary in terms of size, temperament, and production. Whether you want to raise cattle for meat, eggs, or decoration, this section will walk you through the process of selecting the best breed for your agricultural objectives.

Establishing a Farm

For the health and production of your geese, you must provide the perfect habitat. This includes erecting appropriate housing, giving grazing areas plenty of room, and making sure there is enough feed and water available. The basic components of establishing a geese farm, such as planning and building suitable shelters and overseeing pasture and water supplies, will be covered in this part.

Everyday Handling and Administration

Successful daily management and care are essential to your geese farm's success. This covers daily duties including feeding, keeping an eye on health, and keeping things tidy. Maintaining a healthy flock and avoiding frequent problems may be achieved by being aware of the needs of geese on a daily basis, including their behavioral habits and health requirements.

Geese Breeding and Raising

Understanding the geese reproductive cycle and caring for young goslings is crucial if you want to breed them. The fundamentals of geese reproduction are covered in this section, including mating, egg incubation, and rearing goslings. Adequate supervision during this phase is essential for the well-being and growth of the juvenile birds.

Well-being and Health

It's critical to keep your geese healthy and happy. Common health problems, preventative strategies, and available treatments are covered in this section. Maintaining a healthy flock and avoiding illness need regular health examinations, immunizations, and biosecurity procedures.

Nutrition and Feeding

Giving geese a healthy diet is essential for their development and output. The nutritional needs of geese are examined in this section, along with their dietary demands at various life phases. It also discusses the many kinds of feed that are accessible and how to create a diet that promotes their general well-being and efficiency.

Selling and Promoting Geese Goods

Once your geese have established themselves, you may think about marketing and selling their meat, eggs, and feathers. This section offers techniques for pricing, branding, and finding prospective clients for selling your own goods. The legal and regulatory aspects of marketing chicken products are also covered.

Troubleshooting and Solving Issues

Like any agricultural endeavor, raising geese has its share of difficulties. This section provides answers to frequently asked questions about anything from handling environmental conditions to handling health difficulties. You can handle and get beyond these obstacles with the aid of some helpful guidance and troubleshooting techniques.

Final Thoughts and Upcoming Projects

In conclusion, learning to farm geese is an interesting and fulfilling endeavor that needs constant supervision and meticulous preparation. You'll be well-equipped to launch and operate a profitable geese farm if you adhere to the rules and suggestions in this book. As you expand your geese farming enterprise, embrace the trip with a sense of excitement and never stop learning and growing.

THE END

www.ingramcontent.com/pod-product-compliance
Lightning Source LLC
Chambersburg PA
CBHW052326220526
45472CB00001B/296